Fatma Megdiche
Choumous Kallel

Evaluation of the demands of the thrombophilia assessment in South Tunisia

Fatma Megdiche
Choumous Kallel

Evaluation of the demands of the thrombophilia assessment in South Tunisia

Thrombophilia profile in venous and arterial thrombosis

ScienciaScripts

Imprint

Any brand names and product names mentioned in this book are subject to trademark, brand or patent protection and are trademarks or registered trademarks of their respective holders. The use of brand names, product names, common names, trade names, product descriptions etc. even without a particular marking in this work is in no way to be construed to mean that such names may be regarded as unrestricted in respect of trademark and brand protection legislation and could thus be used by anyone.

Cover image: www.ingimage.com

This book is a translation from the original published under ISBN 978-3-8416-6068-8.

Publisher:
Sciencia Scripts
is a trademark of
Dodo Books Indian Ocean Ltd. and OmniScriptum S.R.L publishing group

120 High Road, East Finchley, London, N2 9ED, United Kingdom
Str. Armeneasca 28/1, office 1, Chisinau MD-2012, Republic of Moldova, Europe
Printed at: see last page
ISBN: 978-620-5-81318-8

Contents

Thrombophilia is a condition of constitutional or acquired hypercoagulability defined by biological abnormalities that predispose to thrombotic events. The incidence of thrombophilia varies according to the specific population evaluated, however, it remains a rare disorder in the general population compared to other classical risk factors for thrombotic events [1].

Indeed, it is an alteration of hemostasis of multigenic and multifactorial origin which is characterised by an immense heterogeneity of its clinical expression.

The indication for the investigation of thrombophilia markers during a thrombotic event depends on the type of thrombotic event. Indeed, thrombophilia markers are well-established risk factors for venous thromboembolic disease (VTE), but their relevance to the occurrence of arterial thrombosis (AT) remains controversial [2].

Thrombophilia testing should not be performed systematically for every thrombotic event. In fact, several scientific societies have detailed the different indications that justify the performance of this assessment [1].

Therefore, in our study we focused on evaluating the requests for thrombophilia assessment reviewed at the hematology laboratory of the Habib Bourguiba University Hospital (CHU) in Sfax, on bringing back the epidemiological, topographical and etiological characteristics of thrombotic events and on detailing the biological profile of thrombophilia through the exploration of the different prothrombotic markers following the occurrence of VTE and BP.

1.1. Thrombophilia

1.1.1. Definition

Thrombophilia is a disorder of hemostasis predisposing to thrombosis. It may be of constitutional origin and may be due either to loss of function of certain coagulation inhibitors (antithrombin (AT) deficiency, protein C (PC) deficiency, protein S (PS) deficiency, active protein C resistance (APCR)), It may be due to a gain of function by increasing the expression of certain pro-coagulant factors through various mutations, or it may be of acquired origin, which is essentially represented by the anti-phospholipid syndrome (APS) [1].

1.1.2. Constitutional Thrombophilia

1.1.2.1.   Reminder of the physiology of coagulation

Under physiological conditions a balance exists between thrombotic and anti-thrombotic mechanisms preventing thrombus formation. This balance is governed by hemostasis, which consists of three stages: primary hemostasis, coagulation and fibrinolysis **(Figure 1).**

Coagulation is a fundamental stage of hemostasis which occurs in three steps: initiation, amplification and propagation and which results in the transformation of fibrinogen into an insoluble substance, fibrin, under the action of thrombin or factor IIa. This fibrin formation is the product of a long enzymatic cascade involving the various coagulation factors.

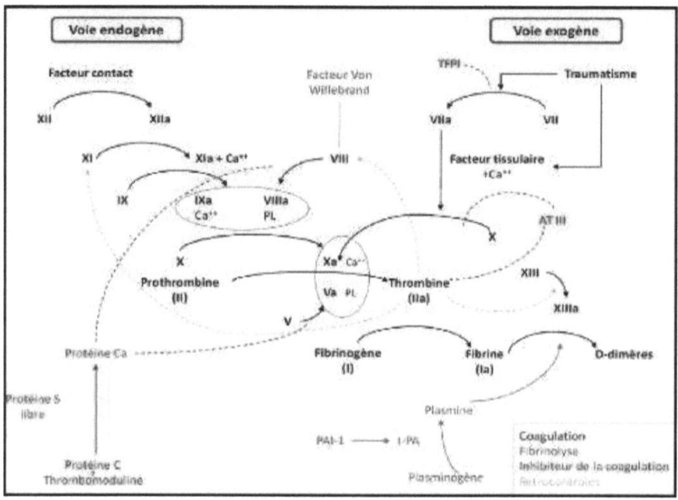

Figure 1: Scheme of hemostasis [3].

Several regulatory systems are involved to ensure that clot formation is rapid, localised and limited to achieve healing without obstructing the vascular lumen, including coagulation inhibitors such as AT, PC and its cofactor PS, the *tissue factor pathway inhibitor* (TFPI) [4].

1.1.2.2. Hereditary S-protein deficiency

❖ **Presentation of the gene**

The *PROS1* gene encoding PS is located on chromosome 3 (3q11.1). The *PROS2* gene, which is a functionally inactive pseudo gene, is homologous to the *PROS1 gene and is* inherited in an autosomal dominant fashion.

❖ **The structure of the S protein**

PS is a vitamin K-dependent glycoprotein.

It contains a gamma-carboxy-glutamic acid domain and an *endothelial growth factor* (EGF) domain. There is also a thrombin-sensitive region and a carboxy-terminal region highly homologous to sex hormone binding *protein* (SHBG).

❖ **The activation of protein S**

PS is present in two circulating forms in plasma: a physiologically inactive *C4-*

binding protein (C4BP) bound form (60%) and a physiologically active free form (40%) that is responsible for anticoagulant activity **[5].**

The PS-C4BP complex is formed by the binding of PS to the β chain of C4BP, which exists in two forms: a form with the β chain (about 10-15%) and a form without this chain.

The dissociation of the PS-C4BP complex giving the active form of PS is dependent on the calcium concentration. The lower the calcium concentration, the greater the dissociation.

❖ **The role of protein S in coagulation**

PS acts as a cofactor of CP in the inactivation of factors Va and VIIIa. It inhibits the activation of prothrombin and the formation of the prothrombinase complex on PLs as well as the activation of factor X.

❖ **Genomic abnormalities of the protein S gene**

Several mutations in the *PROS1* gene have been described. This heterogeneity has made it difficult to establish a causal link between a well-identified mutation and clinical manifestations. The inheritance of this deficit is autosomal dominant.

❖ **The different types of S-protein deficiency**

There are three types of deficits: type I and type III are quantitative deficits (for type I the total PS rate is decreased, while for type III this rate is normal and the free PS rate is decreased), while type II is a qualitative deficit (the PS activity is decreased).

These three phenotypes were described on the basis of total protein, PS antigen concentration, free PS concentration and PS functional activity.

1.1.2.3. Hereditary protein C deficiency

❖ **Presentation of the gene**

The CP gene, *PROC*, is 11.6 kb long and located on chromosome 2q13-14, consisting of nine exons and eight introns **[6].**

PC deficiency is inherited in two ways: autosomal dominant giving the classic form of the deficiency, and autosomal recessive giving the more severe form.

❖ **The structure of protein C**

CP is a zymogen of a vitamin K-dependent serine protease. CP is a glycoprotein synthesised in hepatocytes and circulates in plasma as a heterodimeric complex consisting of a heavy chain containing the factor IIa cleavage site and a catalytic site and a light chain containing the phospholipid (PL) binding site [7].

❖ **Protein C activation**

Physiological activation of CP occurs on the surface of endothelial cells and requires two membrane receptors: "*endothelial protein c receptor* (EPCR) and thrombomodulin. The latter are necessary cofactors for the conversion of CP to active CP (PCa) mediated by factor IIa.

The binding of thrombomodulin to factor IIa forming a complex amplifies CP activation 3-4 times, while the binding of EPCR to zymogen increases the rate of CP activation by a factor of 20 [7].

❖ **The role of protein C in coagulation**

PCa, in the presence of its cofactor, PS, as well as calcium and phospholipids (PL), inhibits the generation of factor IIa by inactivating factors Va and VIIIa [6].

Alternatively, PCa can function as an indirect fibrinolytic agent. It binds to and neutralises *plasminogen activator inhibitor-1* (PAI-1*)*, thereby increasing plasminogen activity.

In addition, due to the reduced generation of thrombin resulting from the inactivation of factors Va and VIIIa, the activation of the plasminogen-fibrin binding inhibitor *thrombin activatable fibrinolysis* inhibitor (TAFI) is reduced, resulting in increased profibrinolytic activity [7, 8].

In addition to its anticoagulant and fibrinolytic function, PCa also has potent cytoprotective and anti-inflammatory properties **(Figure 2)**.

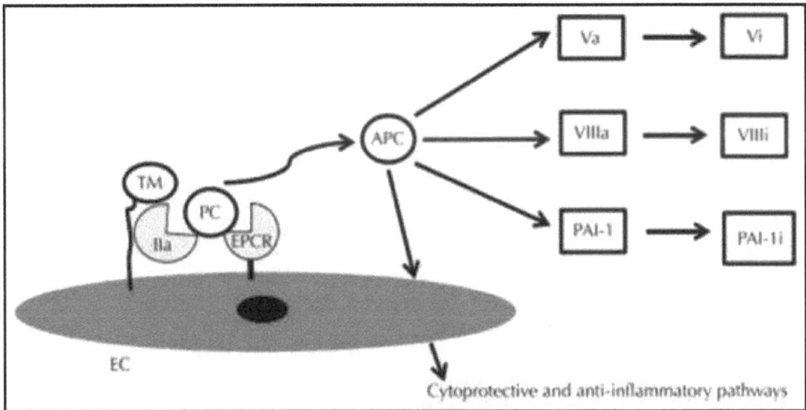

Figure 2: The biological roles of active protein C [7].

❖ **Genomic abnormalities of the protein C gene**

More than 160 mutations in *PROC* responsible for PC deficiency have been described[7] However, underlying mutations with premature stop codons can also be associated with small deletions or insertions of one or a few nucleotides.

The homozygous profile of PC deficiency is rare while the heterozygous profile is somewhat more frequent.

❖ **The different types of protein C deficiency**

There are two types of deficit in PC:

❖ The quantitative deficit type I is characterized by a simultaneous reduction in concentration and activity and is the most frequent deficit;

❖ The type II quantitative deficit is characterised by normal or high concentration with reduced activity [9].

1.1.2.4. Hereditary antithrombin deficiency

❖ **Presentation of the gene**

The AT gene, *SERPINC1,* is located on chromosome 1q23-25. It has seven exons covering 13.4 kb of base pairs. It is inherited in an autosomal dominant pattern.

❖ **The structure of antithrombin**

AT is a single-stranded glycoprotein of 432 amino acids with three disulphide bridges and four N-glycosylation sites on the asparagines. Position 135, which is only partially glycosylated, accounts for the two glycoforms of AT found in plasma: a (90% of AT) and в (10%), with four and three N-glycans respectively [10]. AT has two main functional sites: a reactive site and a heparin binding domain.

❖ **Antithrombin activation**

It is an allosteric activation of AT by heparin derivatives leading to an optimal exposure of its reactive site

❖ **Role of antithrombin in coagulation**

AT is an inhibitor of serine proteases (serpins) which mainly inactivates active factors generated by the coagulation cascade (Figure 3), including factors IIa, Xa and IXa and, to a lesser extent, factors XIa and XIIa, as well as kallikrein and plasmin [11]. It also plays a role in the inactivation of extrinsic tenase (tissue factor IIa). AT interacts with the active site of coagulation serine proteases by forming a complex with the latter.

In the absence of heparin, AT inhibits coagulation proteases in a progressive manner. When heparin binds to the AT binding site, AT undergoes a conformational change that improves its inhibitory activity by more than 1000-fold.

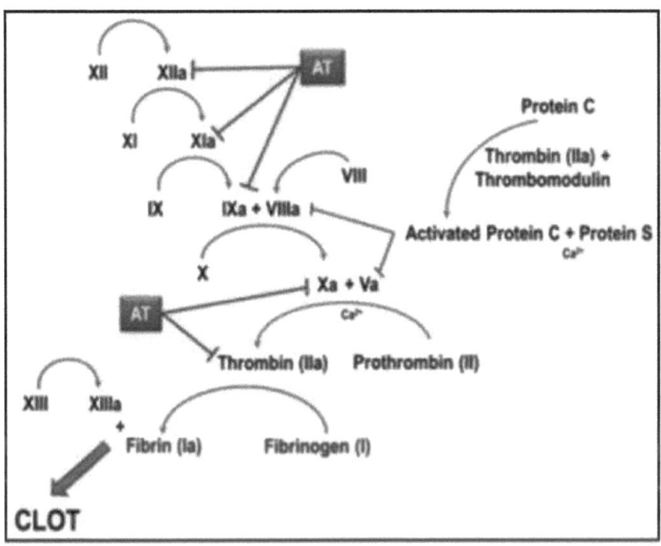

Figure 3: Rôle de l'antithrombine dans la coagulation [11]

❖ **Genomic abnormalities of the antithrombin gene**

At the level of the endoplasmic reticulum 1 AT undergoes two significant post-translational modifications: three intramolecular disulfide bridges and N-glycosylation. Ineffective glycosylation at position 135 accounts for the presence of two glycoforms: a-AT, with four N-glycans, which is the major glycoform in plasma, and в-AT with three N-glycans which has more affinity to heparin **[12]**. Three different mutations affecting the glycosylation consensus sequence (N135-K136-S137) have a direct effect on two functional characteristics of AT: affinity for heparin and inhibition of factor Xa.

❖ **Impact on antithrombin activity**

Allosteric activation of AT causes the appearance of an exosite which is involved in the inhibition of factors Xa and IXa. The inhibition of factor IIa depends essentially on the matrix effect of long-chain heparin. Indeed, by acting as a matrix, heparin allows factor IIa to be brought closer to AT with a favourable orientation of AT in the catalytic groove of factor IIa.

❖ **The different types of antithrombin deficiency**

AT deficiency is either type I or type II (qualitative deficiency). Type I deficiency is most often caused by missense mutations, frameshifts, deletions, nonsense mutations and insertions (of less than 30 base pairs). This is the most frequent deficit and is characterised by a reduction in AT concentration while retaining its activity. Depending on the location of the mutations in the AT gene, *SERPINC1*, the type II deficiency resulting in an abnormal protein can be subdivided into type II- RS *"reactive site"*, type II-HBS *"heparin binding site"* and type II-PE *"pleiotropic effect"* [11]. The type II-RS deficiency is characterised by a marked thrombogenic effect, whereas the type II-HBS deficiency has little or no thrombogenic effect [13].

1.1.2.5. Active protein C resistance and the G1691A gene polymorphism

❖ **Presentation of the gene**

The gene encoding factor V is located on chromosome 1q23[14].

❖ **The structure of factor V**

Factor V is a 330 kDa glycoprotein with regions homologous to factor VIII in its structure [14].

❖ **Factor V activation**

Activation of factor V to factor Va is achieved by thrombin and/or factor Xa.

❖ **The role of factor V in coagulation**

Factor Va is a cofactor for factor Xa. In the presence of PL and calcium, the factor Va - factor Xa prothrombinase complex leads to the generation of factor IIa and the activation of the coagulation cascade [15].

❖ **Genomic abnormalities of the factor V Leiden (FVL) gene**

This is a missense mutation in factor V, the guanine at position 1691 is replaced by an adenine, the arginine at position 506 is replaced by a glutamine resulting in FVL [16].

❖ **Impact on factor V activity**

Since the arginine at position 506 serves as a cleavage site for PCa, this mutation prevents the degradation of factor V, giving rise to RPCa.

By remaining active for a longer period of time, LVF leads to an increase in thrombin production resulting in the generation of a fibrin excess and thus an increased risk of thrombosis [16].

❖ **Other factor V mutations**

FVL is the most common mutation giving CPPa, however, other mutations exist such as the Cambridge mutation, the HR2 haplotype, the Hong Kong mutation, and the Nara mutation [17, 18].

1.1.3. Acquired thrombophilia

1.1.3.1.   Anti-phospholipid syndrome

APL is an acquired autoimmune thrombophilia characterised by the presence of anti-phospholipids (APL) and the occurrence of thrombotic events and/or obstetric complications. There are two types of APL: conventional APL (lupus anticoagulant, anticardiolipin and anti-e2-glycoproteinI (02GPI) and non-conventional APL (antiphosphatidyl ethanolamine, anti-prothrombin and anti annexin):

✓ **Lupus anticoagulant (LA),** also known as anti-prothrombinase antibodies, are antibodies defined by their ability to prolong, in vitro, PL-dependent coagulation tests. As clinical thromboembolic events are strongly correlated with the presence of LA, it is considered the most important risk factor (RF) for thrombotic and obstetric complications [19]. LA presents a heterogeneous group of antibodies that may or may not depend on the presence of plasma cofactors such as factor II and e2GPI [20]. However, e2GPI-dependent LAs confer a higher risk of thrombosis compared to factor II-dependent LAs [19].

✓ **Anti-cardiolipin antibodies (aCL)** Depending on their dependence on the presence of a plasma cofactor which is e2GPI, there are three types of aCL:

✓ The most pathogenic aCLs found in SAPL recognize domain 1 of e2GPI;

✓ The aCLs that recognise the other domains of the e2GPI are not known to have pathological significance;

✓ aCLs that recognise cardiolipin independently of the e2GPI cofactor are known as true aCLs and are encountered in infections.

■ **Anti-e2GPI antibodies:** e2GPI is organised into five domains, it has two different conformations: a free circular form and an open form bound to anionic PLs via domain 5. In its open form, it exposes its highly antigenic domain I epitopes, allowing the binding of anti-e2GPI **[21]**.

1.2. Thrombotic events

1.2.1. Venous thromboembolic disease

VTE is a complex multifactorial disease that represents a public health problem. It is a chronic and recurrent condition **[22]**.

It consists of partial or total venous obstruction by an endoluminal thrombus, which may be located throughout the venous tree, with the lower limbs being the main site **[23]**.

VTE includes deep vein thrombosis (DVT), the major complication of which is pulmonary embolism (PE).

The mechanisms responsible for the occurrence of VTE are described by the Virchow triad, which combines blood stasis, endothelial wall lesion and alteration of hemostatic balance **[24]** .

1.2.2. Arterial thrombosis

TA is the formation of a clot in an artery that will cause the downstream blood flow to be interrupted.

It can affect different territories such as the coronary arteries leading to myocardial infarction (MI) or the arteries of the brain leading to stroke.

2.1.Patients

This is a retrospective study conducted over a period of 5 years, from January 2013 to December 2017. The study included all patients, aged 18 to 60 years, who benefited from a thrombophilia check-up performed at the hematology laboratory of CHU Habib Bourguiba of Sfax. This study included 1180 thrombophilia assessments.

We have excluded from our study :

- Assessments where the age of the patient is not specified.

- Assessment of cirrhotic patients.

Thrombophilia tests that were performed for one of the following indications: retinal vein occlusion and superficial thrombophlebitis.

- Repeated check-ups of the order of 20 thrombophilia check-ups.

2.2.Methods

2.2.1. Clinical Information Sheet (CIS)

Following a standardised data collection form, we collected data for each patient included in our study.

- Epidemiological data: age, gender, file number, department, origin, occupation, and telephone number

- The main diagnosis : DVT, PE or BP

- Risk factors :   sedentariness/immobilisation, dyslipoproteinemia, diabetes, hypertension, smoking, pregnancy, estrogen/progestin use, venous disease, renal insufficiency (RI) and cirrhosis.

- Personal pathological antecedents (ATCD)

- The topography of VTE: DVT of the IM, DVT of the upper limb (DVT-MS), thrombosis of the portal vein (DVT), of the supra-hepatic vein (TV-SH), of the mesenteric vein (MVM), of the renal vein (TVR), of the cerebral vein (TVC).

BP topography: MI and stroke.

- Personal history of thrombosis: number of thrombosis episodes, age of first episode.

- Family history of thrombosis.

- Anticoagulant treatment at the time of sampling: Heparin, anti vitamin K (AVK).

- Biological data: prothrombin time (PT), active partial thromboplastin time (aPTT), fibrinogen, PS, PC, AT, RPCa, FVL and circulating anticoagulant (CA) type-lupus anticoagulant (LA)

2.2.2. Biological sample

The blood sample was taken in tubes containing 0.109M tri-sodium citrate with a ratio of 1 volume of citrate solution to 9 volumes of blood. The plasma obtained after double centrifugation of these tubes at 3500g for 15 minutes will be frozen.

During the assay, the plasma is thawed and heated to 37°C in a water bath for at least 15 minutes, then homogenised by turning the tubes over a few times.

The determinations were carried out on STA Compact®.

2.2.3. Determination of different physiological coagulation inhibitors

2.2.3.1.   Protein S assay

The determination of PS was carried out with the STAGO STACLOT® protein S kit. This is a time-based determination of PS activity.

As the cofactor of PCa, PS accelerates the rate of inactivation of factors VIIIa and Va, the principle of this test is based on the measurement of the clotting time of a factor Va-enriched system.

The plasma level of functional PS is between 55 and 140%.

2.2.3.2.   Protein C assay

The STAGO STACLOT® Protein C kit is used for the determination of CP. It is a time-controlled determination of functional CP by cephalin time extension and activator.

The principle of this test is based, firstly, on the activation of PC to PCa by an extract of snake venom *(Agkistrodon c.controtrix)*, Protac. Then, in a second step, on the measurement of the TCA prolongation linked to the degradation of factors Villa and Va by PCa. The PC to be measured is provided by the patient's plasma, whereas all other factors are provided in excess by the reagent.

The plasma level of functional CP is between 70 and 130%.

2.2.3.3.   Antithrombin assay

The determination of AT is performed with STAGO STACHROM® antithrombin III reagent. It is a quantitative colorimetric assay by the amidolytic method on a synthetic chromogenic substrate.

The principle of this test is based on the proportionality between the amount of neutralised thrombin and the amount of AT present in the medium. For this purpose, plasma is incubated in the presence of heparin and a fixed and excess amount of thrombin, then the residual AT is measured by its amidolytic activity on the chromogenic substrate, which is the paranitroaniline group dosed at 405nm.

The plasma level of TA is between 80 and 120%.

2.2.3.4.   Determination of the CPRa

The detection of CPPa is performed by the STAGO STA-Staclot® *activated-protein C-resistance* kit.

Because of its ability to inactivate factor Va, PCa induces a prolongation of the clotting time of plasma in a calcium medium. Any abnormally short extension of this time indicates the presence of PCa.

Coagulation of the diluted sample is performed in the presence of factor V-deficient plasma and a factor X activator (Critalus viridis helleri venom), whose role is to initiate coagulation at this level and eliminate the interaction of upstream factors.

For any clotting time measurement<130 seconds, an RPCa is demonstrated.

2.2.3.5.   Search for the G1691A polymorphism in FVL

The G1691A polymorphism in FVL was detected by the PCR-AS (polymerase chain reaction-allele specific) method. The principle of this method is based on the fact that the presence of a mismatch at the 3' end of a primer does not allow amplification. Two amplifications are performed in two tubes, both containing a common primer, with a wild-type primer added to the first tube and a specific primer added to the second tube. DNA extraction is performed from whole blood collected on an EDTA tube by the saline method. The amplified PCR products are visualised by electrophoresis after migration on a 2% agarose gel. Three profiles are found: healthy subject (no mutation), homozygous profile and heterozygous profile.

2.2.3.6. Searching for LA type CCAs

The investigation of LA type ACC is based on four steps according to the recommendations of the *International Society of Thrombosis and Haemostasis* (ISTH).

Step 1: Screening tests

- Sensitised APC: (with low PL concentration and silica as activator) this sensitisation allows to accentuate the APL prolongation. An APTT is considered normal if the ratio of sick time (M time)/control time (T time) is strictly less than 1.2 seconds.

- *"dilute Russell's Viper Venom Time*: Russell's Viper Venom is a
The factor X activator, which in the presence of PLs and calcium, triggers coagulation at this level and eliminates the interaction of factors upstream in the coagulation cascade.

Step 2: Demonstration of the inhibitory effect

- The interpretation of the result obtained is done by calculating the Rosner index [28]: IR= [Time (T+M)-time [2]/time [2]] x 100

- The presence of ACC type LA is confirmed if IR >= 15%.

Step 3: Confirmatory tests

The screening tests are repeated in the presence of a high concentration of PL to demonstrate the phospholipid-dependent nature of the detected inhibitor.

Step 4: Exclusion of other associated coagulopathies (anti-VIII, heparin and endogenous coagulation factor deficiency).

The exploration of aCL and anti-e2GPI is done by the ELISA immunological test (ORGENTEC® DIAGNOSTIKA kit). This technique aims to detect the IgG isotypes of these antibodies.

A positive result must be confirmed on a repeat sample taken 12 weeks apart. Normal values are less than 10 units/mL.

2.2.4. Statistical study

Data entry and statistical analysis were carried out using SPSS version 20 software. For qualitative variables, simple and relative frequencies were calculated. Means, standard deviations, medians and range (extreme values) were calculated for the quantitative variables.

During the period from 1er January 2013 to 31 December 2017, 1180 thrombophilia tests were collected in the hematology laboratory of the Habib Bourguiba Hospital in Sfax, of which 969 (82.1%) tests were received for VTE and 211 (17.9%) tests were received for TA **(Figure 4)**

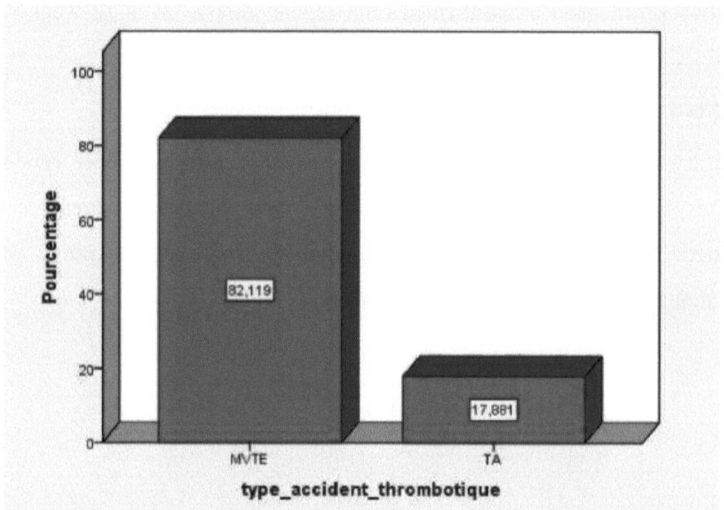

Figure 4: Percentage of collated assessments by type of thrombotic event

In the following, the results will be presented according to the type of thrombotic event observed.

2.3. Venous thromboembolic disease

2.3.1. Epidemiological characteristics

2.3.1.1. Number of assessments collected per year

In our retrospective study, conducted over a period of 5 years, we collected 969 requests for thrombophilia tests prescribed as part of an etiological investigation for young adults with at least one episode of VTE. The number of such requests during the study period ranged from 172 to 210 requests/year **(Figure 5).**

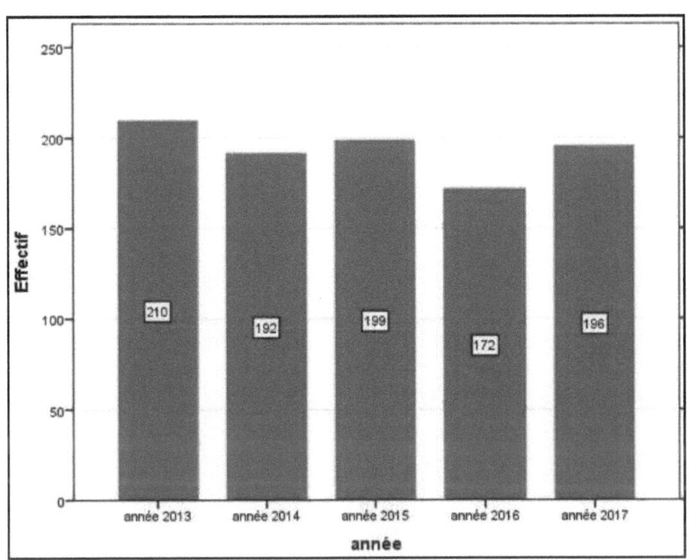

Figure 5: Number of thrombophilia check-ups performed for venous thromboembolic disease per year

2.3.1.2.   Gender distribution

Of the 969 patients, 555 were female (57.28%) and 414 were male (42.72%) with a sex ratio (M/F) of 0.74.

2.3.1.3.   Age distribution

In our series, the mean age was 40.64 ±11.34 years with extremes ranging from 18 to 60 years. **Figure 6** shows the distribution of patients by sex and age group.

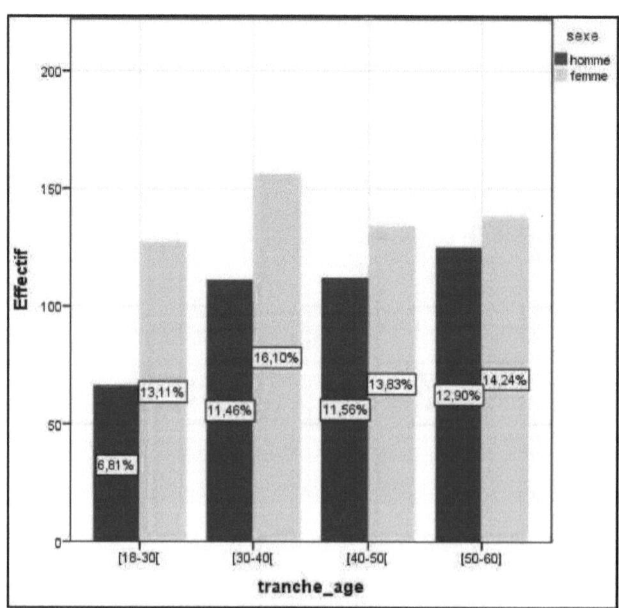

Figure 6: Distribution of patients with venous thromboembolic disease by sex and age group

2.3.1.4. Breakdown by department requesting the review

The service requesting the review was specified in 844 requests, i.e. 87.1%.

The majority of requests came from the internal medicine department with 523 requests (54%), followed by the neurology department with 104 requests (10.7%), the intensive care department with 36 requests (3.7%) and the pneumology department with 30 requests (3.1%) **(Figure 7)**.

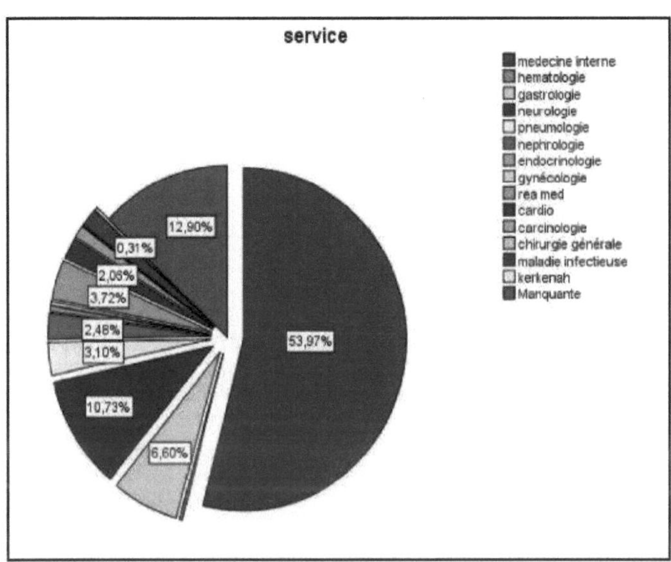

Figure 7: Distribution of thrombophilia check-ups performed in the context of venous thromboembolic disease according to the requesting department

2.3.2. Topographical features of venous thromboembolic disease

The type of thromboembolic event was mentioned in 732 reports (76.3%). These were 609 cases of DVT and 123 cases of PE.

2.3.2.1.   Deep vein thrombosis

The 609 DVT cases were distributed as follows: 206 cases of IM DVT (21.2%), 191 cases of non-specific DVT (19.7%), and 212 cases of non-specific DVT (21.8%).

* **DVT of the IM:**

DVT of the IM was observed in 206 patients or 21.2% with a sex ratio M/F 1.23.

* **Unusual seat TVP**

These were DVTs of unusual location, in our series we noted MS-TV, p-TV, SH-TV, RST, CT and MST **(Table I).**

Table I: Topography of DVTs of unusual location

Seat	Workforce	Percentage	Sex ratio M/F
TV-MS	47	4,8	1,04
TVp	42	4,3	0,75
TV-SH	7	0,7	0,16
TVR	7	0,7	2,5
TVC	81	8,3	0,5
TVM	7	0,7	2,5

2.3.2.2. Non-specific DVT

DVT-ns was encountered in 212 patients or 21.8% with a sex ratio of 0.89 M/F.

2.3.2.3. Pulmonary embolism

PE was found in 123 patients (12.6%) with a sex ratio of 0.64:1.

Twenty-seven patients had PE associated with DVT.

2.3.3. Personal and family history

2.3.3.1. Personal history

❖ **History of VTE :**

In our series, 90 patients had a history of VTE, i.e. 9.3%, with a predominance of males (sex ratio 1.05).

The number of VTE episodes varies between 1 and 5 episodes **(Table II).**

Table II: Number of episodes of venous thromboembolic disease by number and percentage of patients

Number of episodes	Number of patients (%)
An episode	55 (5,7%)
Two episodes	18 (1,9%)
Three episodes	10 (11,1%)
Four episodes	5(0,5%)
Five episodes	2(0,2%)

The age of onset of the first episode of VTE ranged from 11 to 56 years with a mean of 38 ± 9.2 years.

❖ **Other Pathological history :**

In our series, 70 patients had a pathological history. **Table III** details the different pathologies found.

Table III: The different pathological antecedents of the patients

Pathology	Number of patients
A systemic disease:	45 (4,6%)
Behcet's disease	10
Still's disease	1
Gougerot-Sjogren's syndrome	8
Systemic lupus erythematosus	26
Obstetrical events :	33 (3,4%)
Postpartum	15
Recurrent miscarriage	15
Pregnancy toxemia	3
Hepatic injury:	6 (0,6%)
Hepatitis B	3
Hepatitis C	1
Hydatid cyst of the liver	2
Cancer:	4(0,4%)
Ovarian tumour	2
Neoplasia of the lung	1
Chronic lymphocytic leukemia	1
Other pathologies encountered	7 (0,7%)
Ischemie digitale	2
Hemorrhagic rectocolitis	2
C^liac disease	2
Rheumatic fever	1

2.3.3.2. Family background

A family history of VTE was found in 28 patients with a frequency of 2.9%. These cases involved first-degree relatives (ascendants, descendants and siblings).

2.3.4. Risk factors for venous thromboembolic disease

2.3.4.1. Classic risk factors

FDRs for VTE were found in 131 patients (13.5%).

A single FDR for VTE was found in 115 patients, while a combination of several FDRs was observed in 16 patients. **Table IV** illustrates the frequency of the different FDRs and their distribution according to sex.

Table IV: Distribution of risk factors for venous thromboembolic disease by gender

FDR	Number (%)	Gender	
		Male	Woman
Dyslipoproteinemia	8 (0,8%)	3	5
Immobilization	61 (6,2%)	28	33
Pregnancy	25 (2,5%)	-	25
Diabetes	10 (1,03%)	1	9
HTA	8 (0,8%)	3	5
IR	12 (1,2%)	6	6
Smoking	3 (0,3%)	2	1
Taking an oestrogen	2 (0,2%)	-	2
Venous terrain	2 (0,2%)	1	1

The distribution of VTE DRFs by 10-year age groups is shown in **Figure 8**, with the first age group showing a peak in pregnancy and the other age groups showing a predominance of immobility.

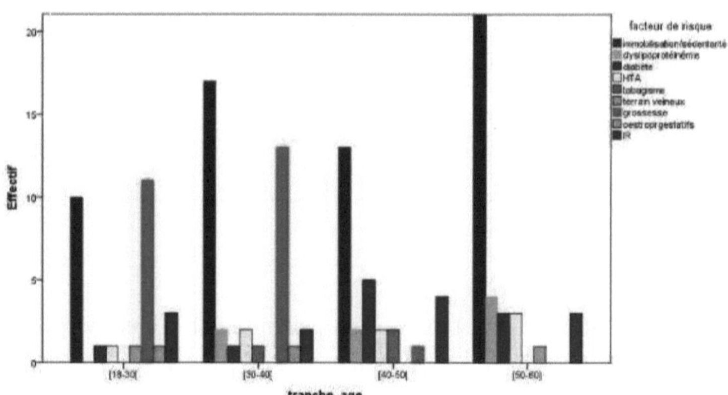

Figure 8: Distribution of risk factors for venous thromboembolic disease by age group.

2.3.4.2. Thrombophilia markers

❖ **Markers of constitutional thrombophilia**

PS deficiency: PS was investigated in 294 reports. A PS deficit was observed in 38 patients (3.9%).

PC deficiency: PC testing was performed in 568 work-ups. CP deficiency was observed in 28 patients (2.8%).

AT deficiency: AT exploration was performed in 925 work-ups. A deficit in AT was found in 44 patients (4.5%).

CPP: CPP was investigated in 890 patients, with CPP found in 138 patients (14.2%).

❖ **FVL polymorphism research :**

The molecular biology results provided in our study only concerned patients from the year 2017. The search for the FVL mutation was carried out for 144 patients, the mutation was detected in 36 patients (3.7%), 23 of whom were men and 13 women.

For the year 2017,36 CPPa were deceased, according to the molecular biology results, the FVL polymorphism was responsible for 36/36 of the CPPa of which 34 patients had a heterozygous mutation and two patients had a homozygous mutation **(Figure 9).**

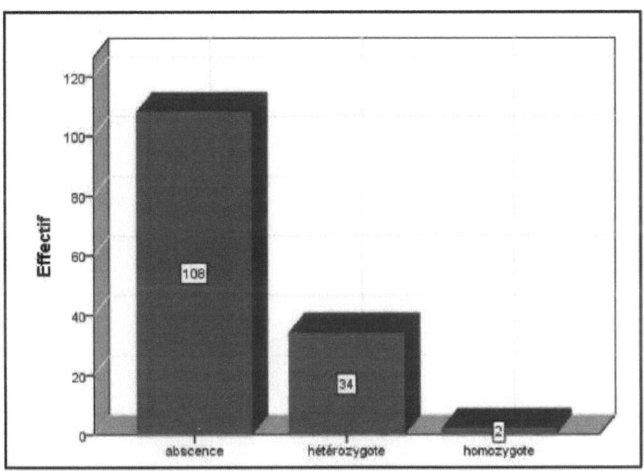

Figure 9: Le polymorphisme du facteur V Leiden chez les patients ayant une maladie veineuse thromboembolique

❖ **Markers of acquired thrombophilia :**

LA-type ACC was the only conventional APL explored in our study. The search for type LA ACC was performed in 862 work-ups, type LA ACC was found in 66 patients.

Of these, six were checked at 12 week intervals and antibody positivity was confirmed in only four patients. Therefore, LA ACC was found in 64 patients or 6.6%.

Table V below summarises the different thrombophilia abnormalities found in our study population, and illustrates the topography of VTE for each abnormality.

**Table V: The distribution of thrombophilia anomalies according to the topography
of venous thromboembolic disease**

	Percentage	TV-EP	TV-TVC	TVR	TVM	%MIMS	TVp	TV-SHns	TV-	TV-
SP deficit	3.9 %	12	3	4	5	1	-	2	-	7
PC deficit	2.8 %	11	2	1	5	1	-	2	-	5
TA deficit	4.5 %	10	7	-	2	14	-	-	-	11
CPRa	14.2 %	53	18	8	4	2	2	-	1	45
FVL polymorphism	3.7 %	10	4	-	4	4	-	2	-	12
ACC type LA	6.6%	20	13	2	4	-	-	1	-	12

2.3.4.3. Risk factors for recurrence of venous thromboembolic disease

The results of the statistical study of the imputability of the different recurrence FDRs in the occurrence of recurrent thromboembolic events are shown in **table VI.**

Table VI: Involvement of risk factors in the occurrence of recurrent venous thromboembolic disease

FDR	Recidivism (+)	Recidivism (-)
Immobilization	6	55
	p=0,741	
Taking oestroprogestins	0	2
Pregnancy	3	22
	*p=0,7*13	
Postpartum	1	14
	p =0,724	
Cancer	0	4
TA deficit	6	50
	p =0,661	
SP deficit	2	42
	p=0,422	
PC deficit	4	32
	p=0,644	
CPRa	18	167
	p=0,573	

FVL mutation	2	0
	p=0,972	
ACC+	10	72
	p =0,361	
Male gender	46	368
	p=,091	
FDR schemes	0	23

We found that these DRFs were not statistically associated with the occurrence of recurrent VTE (p > 0.05). Thus, the imputability of these DRFs in the occurrence of recurrence of MVTE was negated in our study.

2.3.5. Distribution of coagulation defects

From 969 collated reports, 274 reports containing biological abnormalities were found in patients with thrombophilia, of which only 24 reports were remotely controlled. The abnormalities found in these patients are shown in **Table VII.**

Table VII: The biological profile of thrombophilia in patients with venous thromboembolic disease

	Anomaly	Number	Percentage % of total
Isolated constitutional anomaly	SP deficit	19	1.9%
	PC deficit	14	1.4%
	TA deficit	35	3.6%
	RPCA	116	12%
	Total	184	**18.9%**
Anomaly Acquired	ACC type LA	50	**5.1%**
Combined constitutional abnormalities	Combined anomaly in PS and PC	8	0.8%
	Combined anomaly in PS and RPCA	8	0,8%
	Combined PC and AT anomaly	2	0.2%
	Combined anomaly in PC and RPCA	1	0,1%
	Combined anomaly in TA and RPCA	5	0.5%
	Total	24	**2,47%**
Mixed constitutional and acquired anomalies	ACC+ and RPCA	8	0,8%
	ACC+ and SP deficit	3	0.3%
	ACC+ and TA deficit	2	0.2%
	ACC+ and PC deficit	1	0.1%

| | Total | 14 | 1.4% |

Isolated constitutional thrombophilia was found in 184 patients or 18.9%, combined constitutional thrombophilia was reported in 24 patients or 2.47%. LA type ACC was found in isolation in 50 patients or 5.1% and in association with a natural coagulation inhibitor deficiency or CPPa in 14 patients, making the percentage of mixed thrombophilia 1.4%.

2.4.Arterial thrombosis

2.4.1. Epidemiological characteristics

2.4.1.1.   Number of assessments collected per year

The number of thrombophilia check-ups performed in the context of BP ranged from 23 to 59 check-ups/year with an average of 42 check-ups/year **(Figure 10)**.

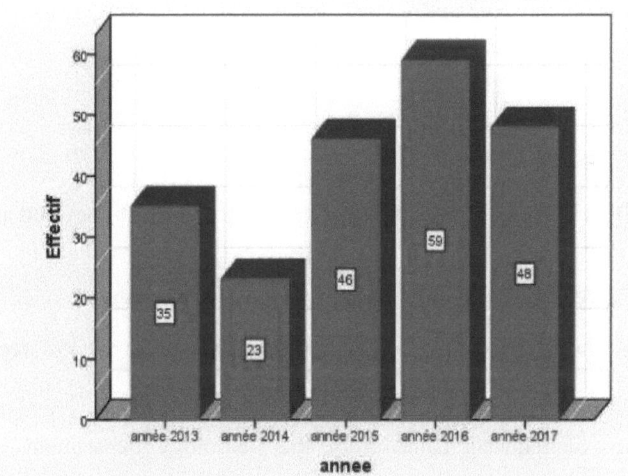

Figure 10: Number of thrombophilia check-ups performed in the context of a TA per year.

2.4.1.2.   Gender distribution

Of the 211 patients, 123 were male (58.2%) and 88 were female (41.7%). There was a predominance of males with a sex ratio (M/F) of 1.39.

2.4.1.3.   Age distribution

The average age was 36.78 +/- 9.56 years, ranging from 18 to 60 years. This distribution shows a predominance of the 30-40 age group representing 40% of the BP population **(Figure 11)**.

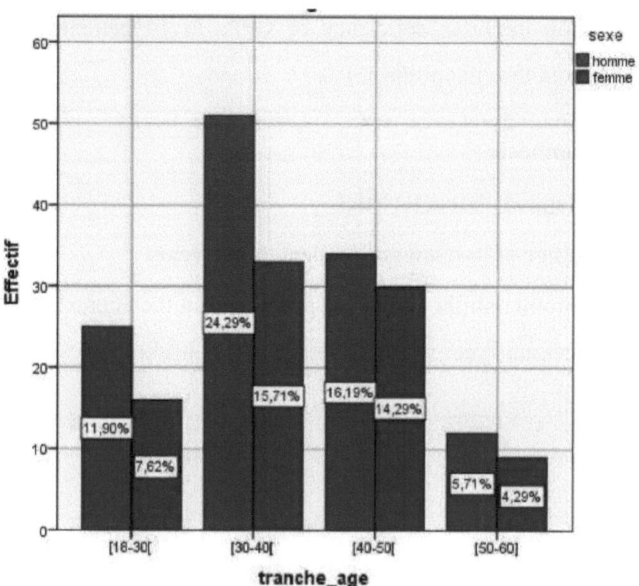

Figure 11: Distribution of patients with arterial thrombosis by sex and age group

2.4.1.4.   Breakdown by department requesting the review

Of the 211 reports, the requesting service was specified in 198 reports, i.e. 93.8%.

The majority of requests came from the neurology department with 118 requests. The department of internal medicine came second with 52 requests. This distribution is illustrated in **figure 12.**

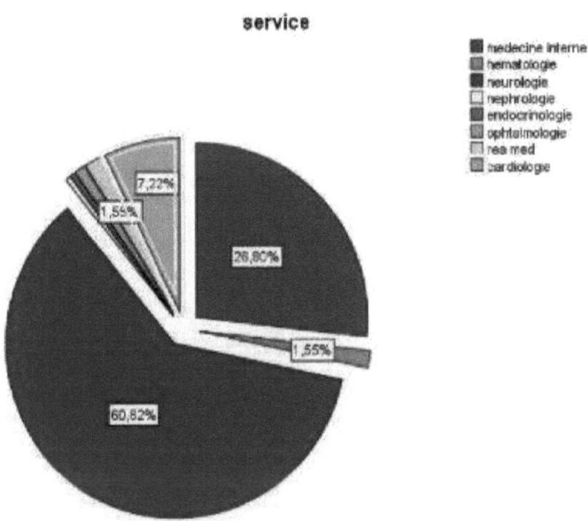

Figure 12: Distribution of thrombophilia tests performed in the context of arterial thrombosis according to the requesting department

2.4.2. Topographical features of arterial thrombosis

BP was predominantly represented by stroke, which was seen in 180 patients (85.3%), MI was seen in 22 patients (10.4%) and non-specific BP was seen in 9 patients (4.2%) **(Figure 13)**.

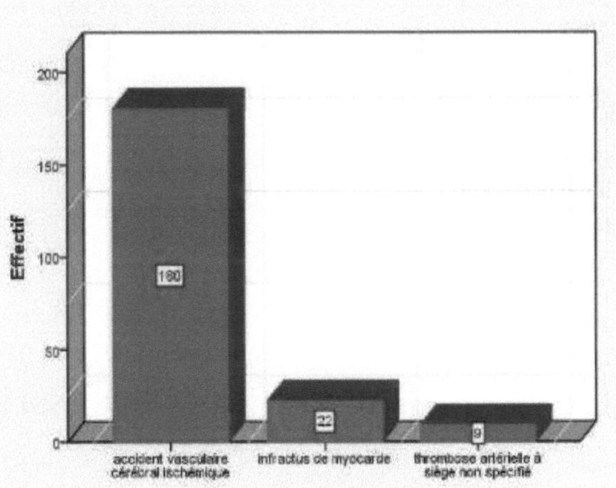

Figure 13: Distribution by type of arterial thrombosis

2.4.3. Personal and family history

2.4.3.1.   Personal history

In our series, 22 patients had a history of BP (10.4%). Other pathological findings were observed in only eight patients. These were obstetric complications in three patients and systemic disease in the other five patients.

2.4.3.2.   Family background

A family history of BP was found in 3 patients with a frequency of 2.4%. These cases involved first-degree relatives (ascendants, descendants and siblings).

2.4.4. Risk factors for arterial thrombosis

2.4.4.1.   Classic risk factors

Based on the completed CR forms, LD FDRs were found in 15 patients or 7.1% (see **Table VIII**). A single LD FDR was found in 13 patients, while a combination of several FDRs was observed in two patients.

Table VIII: Distribution of risk factors for arterial thrombosis by gender

FDR	Number (%)	Gender	
		Male	Woman
Dyslipoproteinemia	5 (2,3)	3	2
Immobilization	1 (0,4)	-	1
Pregnancy	0	-	-
Diabetes	2 (0,9)	1	1
HTA	4 (1,8)	1	3
IR	0	-	-
Smoking	2 (0,9)	2	-
Taking an oestrogen	1 (0,4)	-	1
Venous terrain	0	-	-

2.4.4.2.   Thrombophilia markers

❖ **Protein S deficiency:**

PS was investigated in 184 reports. A PS deficit was observed in 11 patients (5.2%).

❖ **Protein C deficiency:**

CP measurement was performed in 185 patients, a deficit was found in five patients or 2.3%.

❖ **Antithrombin deficiency :**

TA was investigated in 205 patients, and a deficit was found in five patients (2.3%).

❖ **The CPPa :**

CPP was detected in 186 patients, and was found in 27 patients (12.7%).

❖ **FVL polymorphism :**

The molecular biology results provided in our study only concerned patients from the year 2017. The FVL mutation was found in twelve patients or 5.6%. A CPP was found in all these patients, of which eight patients had a heterozygous mutation and four patients had a homozygous mutation.

❖ **Typical CCA LA :**

The search for LA type ACC was carried out in 197 patients and was found in 31 patients (14.6%).

The distribution of thrombophilia abnormalities according to BP topography in our study population is shown in **Table IX.**

Table IX: The distribution of thrombophilia anomalies according to the topography of the arterial thrombosis

Thrombophilia anomaly	Percentage of anomalies	STROKE	IDM	Non-specific arterial thrombosis
SP deficit	5.2%	10	1	-
PC deficit	2.3%	5	-	-
TA deficit	2.3%	3	1	1
CPRa	12.7%	20	5	2
FVL polymorphism	5.6%	4	8	-
ACC+	14.6%	17	8	6

2.4.5. Distribution of coagulation defects

Isolated constitutional thrombophilia was found in 31 patients or 14.6%, combined constitutional thrombophilia was found in four patients or 1.8%.

LA ACC was reported in isolation in 24 patients (11.3%) and in association with a constitutional defect in seven patients (3.3%).

Table X details the biological profile of thrombophilia in patients with BP.

Table X: Biological profile of thrombophilia in patients with arterial thrombosis

	Anomaly	Number	Percentage % of total
Isolated constitutional anomaly	SP deficit	6	2.8%
	PC deficit	3	1.4%
	TA deficit	2	0.9%
	RPCA	20	9.4%
	Total	**31**	**14.6%**
Anomaly Acquired	ACC type LA	24	11.3%
Combined constitutional abnormalities	Combined anomaly in PS and PC	1	0.4%
	Combined anomaly in PS and RPCA	1	0.4%
	Combined anomaly in PS and AT	1	0.4%
	Combined anomaly in PS, PC, and AT	1	0.4%
	Total	4	1.8%
Mixed constitutional and acquired anomalies	ACC+ and RPCA	5	2.3%
	ACC+ and SP deficit	1	0.4%
	ACC+ , TA deficit and CPPa	1	0.4%
	Total	7	3.3%

2.5. Qualitative aspect of requests for thrombophilia assessment

2.5.1. Evaluation of clinical information sheets

2.5.1.1.   In the context of venous thromboembolic disease

Among the 969 requests for thrombophilia tests that were received by the hematology laboratory in the context of VTE, there were several missing data.

Indeed, the department requesting the thrombophilia assessment was not specified in 125 records (12.8%). In addition, no clinical information was provided in 112 requests (11.5%)   : the clinical characteristics of the patient were not specified.

In addition, the epidemiological characteristics of the patients, their personal and/or family history of thrombosis and their therapeutic behaviour were missing, and 288 requests for analysis were found to be without indication (29.7%).

2.5.1.2.   In the context of arterial thrombosis

For the 211 requests collected in the framework of a TA, the service requesting the thrombophilia assessment was not specified in 13 records (6.11%). In addition, no clinical information was provided in 57 requests (27%).

This lack of information has led to an underestimation of pro-thrombotic factors. This leads us to increase the awareness of prescribers to provide sufficient information in their requests to guarantee the quality of the results.

2.5.2. Evaluation of the relevance of requests for thrombophilia check-ups

Referring to the GEHT and SFMV recommendations and the latest 2019 recommendations from French learned societies [1] detailing the indications for thrombophilia work-up, compliance was assessed in our Vindication of thrombophilia work-up in VTE study series.

In fact, 423 applications were justified (43.6%) and distributed as follows:

- Unprovoked first episode of proximal DVT or PE: 67 cases

- Induced or uninduced recurrence of proximal DVT or PE: two cases

- Recurrence of unprovoked distal DVT: 77 cases

- Induced or uninduced first episode of proximal DVT or PE in women of childbearing age: 57 subjects

- Unusual seat thrombosis: 212 subjects

- DVT or PE associated with skin necrosis or with a history of placental vascular pathology: two cases

- DVT or PE with a family history of first degree thrombosis: 9 cases

Our results were similar to those of the French study by Allain et al. who found that 57% of the tests performed were in accordance with the recommendations [25] and the American study by Kwang et al. who found that the compliance of the tests with the GEHT and SFMV recommendations was around 56.6% [26]. As regards thrombophilia tests, in line with our results, their indication was considered inappropriate according to the literature. This leads us to recall the indications for performing a thrombophilia work-up in the context of VTE in order to obtain more relevant and targeted biological investigations. On the other hand, the indication for thrombophilia testing in the context of arterial thrombosis remains controversial in the literature, several studies have questioned the usefulness of this test as it is not clinically useful or cost-effective [2], however, several studies justify the performance of this test in subjects aged less than 50 years [27].

In our study, of the 211 patients with arterial thrombosis, 190 patients were under 50 years of age, i.e. 90%.

2.5.3. Assessment of the completeness of the balance sheets

Considering that our work was etiological, a thrombophilia work-up was considered complete if the determination of PS, PC and AT, the search for a

CPPa and the search for LA type CCA were completed.

Of the thrombophilia tests received by the laboratory in the context of VTE, 108 were complete, i.e. 11.1%.

In addition, a thrombophilia work-up was considered to be controlled if the thrombophilia markers PS, PC, AT, and CPPa were repeated on a second sample before concluding a natural anticoagulant deficiency or CPPa. In the case of APS, antibody positivity can only be demonstrated after a repeat test at 12 week intervals.

In our study, only 25 check-ups were controlled, i.e. 2.5% (see Appendix 1), of which six controls concerned patients with LPA, which is a much lower result compared to the results of other series where the percentage of controls was 11.9% in the study by Devigne et al [28] and 48.8% in the study by Journaud et al [29].

On the other hand, of the 211 thrombophilia tests prescribed in the context of a TA, only twenty tests were complete, i.e. 9.47%, and seven results were checked remotely on a second sample, i.e. 3.13% (see Appendix 2), of which six checks concerned the CPPa and only one check concerned the SP deficit. In general, of all the tests collected in our study, 128 tests were complete, i.e. 10.8%, and 32 tests were controlled, i.e. 2.7%. The large number of incomplete investigations as well as the absence of controls meant that thrombophilia in our series was probably underestimated.

This leads us to alert clinicians to the importance of performing a complete investigation of thrombophilia markers and the need for subsequent monitoring of the results provided.

2.6. Study of the epidemiological data of our patients

2.6.1. Age

2.6.1.1.   Patients with venous thromboembolic disease In our series, all our patients were in the age range of 18-60 years. The paediatric population was excluded from our study because of its particularities, as several studies have described the diagnosis, the different indications for thrombophilia work-up and the management of this population [30]. Patients over 60 years of age were also excluded from our study because of the specificities of this population and the fact that biological investigation for thrombophilia is not recommended for a first episode of VTE over 60 years [31].

The average age of our patients was 40.6 ± 11.3 years. The distribution of cases according to ten-year age groups showed two peaks in frequency between 30 and 40 years (27.5%) and between 50 and 60 years (27.41%). Our results are similar to those found in the Tunisian study by Ben Salah et al [32] which showed an average age of 45.7 years and also found a peak in the 30-40 year age group (21.3%).

2.6.1.2.   Patients with arterial thrombosis

The average age of the patients with BP was 36.7 with a peak in the 30-40 age group, which represented 40% of our study population.

Our result is in agreement with the Tunisian study of Kefi et al [33], which showed a mean age of 35.7 years.

2.6.2. Gender

2.6.2.1.   Patients with venous thromboembolic disease Our results

showed a predominance of females with a sex ratio of 0.74. This is consistent with the Tunisian study by Kechida et al. This is in agreement with the Tunisian study by Kechida et al [34] which found a female predominance with a sex ratio of 0.84 and the UK study by Martinez et al which found a sex ratio of 0.81 [35].

Contrary to our result, a male predominance has been noted in other studies with a sex ratio varying between 1.23 [36] and 1.28 [32].

Indeed, the peak in frequency noted in the 30-40 age group with a predominance of women is explained in several studies by pregnancy and the use of oral contraceptives in women at this age, which are among the DRFs triggering VTE [35].

2.6.2.2.   Patients with BP

A predominance of males was found in our patients with BP, which was found in the study by Konin et al. which showed a sex ratio (M/F) of 1.6 **[37]**.

2.6.3. Family background

2.6.3.1.   Patients with venous thromboembolic disease Only 2.9% of these patients had a family history of VTE. This frequency is significantly lower than those found in other studies where they ranged from 7.4% **[38]** to 10.4% **[33]**.

2.6.3.2.   Patients with arterial thrombosis

A family history of arterial thrombosis was found in three patients with a frequency of 2.4%.

Our result was inferior to that found in the Tunisian study by Kefi et al **[39]** where a family history was noted in 10.8% of patients.

Taking into account the large number of patients included in our study, the small number of patients with a family history of thrombotic events suggests a lack of collection of CRs during the interview and requires a greater awareness among prescribers of the need to conduct a detailed and complete interview.

2.7. Evaluation of the results of constitutional and acquired thrombophilia in the occurrence of thrombotic events in the young subject

2.7.1. Importance of pre-analytical conditions in the interpretation of the thrombophilia test

In the medical laboratory, the pre-analytical phase encompasses all actions and aspects of the diagnostic procedure that occur prior to the analytical phase **[40]**, such as patient interviewing, biological sample collection, storage and up to the time of analysis.

Indeed, this phase will have a direct impact on the reliability and validity of the results obtained in the analytical phase, and its standardisation is of crucial importance which has a direct effect on the quality of the results **[41]**.

A comprehensive interview including clinical information is a key element of this phase and is of major importance. Indeed, this is one of the most recent recommendations of the Groupe Francais d'Etude sur l'Hemostase et la Thrombose (GFEHT) and the *International Society on Thrombosis and Haemostasis* (ISTH*)* regarding the proper conduct of the pre-analytical phase [41].

Indeed, in our study, the thrombophilia referral forms filled in by clinicians were our only reference where, despite efforts to sensitise prescribers, we frequently encountered several parts of these forms not filled in and not containing any CR. Using the patients' medical records as a reference, we could have provided complete clinical profiles for our patients. However, the large size of our study population calls into question the feasibility of this alternative. However, cooperation between the clinician and the biologist remains the best solution to obtain the complete patient profile and to make the diagnosis.

On the other hand, a thrombophilia test is a sensitive biological investigation that can be influenced by various factors and whose quality determines the credibility of the results obtained. Indeed, the sample must be taken at a distance from thrombotic events and inflammatory episodes. Anticoagulant treatment, hormonal contraception and any other treatment or pathological situation that may influence the markers of thrombophilia should be reported to the biologist [42].

In other words, when testing for thrombophilia markers, an abnormality can only be confirmed after remote control on a second sample [43], and this control step is often neglected.

This negligence had a direct effect on our results. Indeed, despite the large number of thrombophilia anomalies that were found, only controlled and confirmed anomalies were taken into account, which gave us a low percentage of thrombophilia in our study population. The lack of control of the results could be explained by :

- Clinicians' ignorance of the importance of this step in confirming and demonstrating the reported anomaly.

- Lack of communication between clinicians and biologists due to lack of computerisation between the different clinical departments and the laboratory.

- The cost of medical transport for patients who live far from the hospital is a cause of non-adherence.

- The medical situation of some patients may prevent them from travelling to the laboratory for a second sample.

2.7.2. Profile of thrombophilia marker abnormalities by type of thrombotic event

2.7.2.1.   In venous thromboembolic disease

Exploration of the etiological profile of VTE has demonstrated the incrimination of thrombophilia abnormalities in the occurrence of this condition in several studies. The frequency of thrombophilia in patients with VTE has been reported in the literature to range from 15.1% [33] to 63.5% [32] (Table XI), which is consistent with our findings. In fact, thrombophilia was identified in 271 of our patients (27.9%).

Table XI: Comparative frequencies of thrombophilia in thromboembolic venous disease in some literature series

	Year of study	thrombophilia	Hereditary thrombophilia	Acquired thrombophilia	Mixed thrombophilia
Mateo et al[44]	1997	16,93%	12,85%	4,08%	0,75%
Ben Salah et al[32].	1996 2010	63,5%	22,6%	19,1%	-
Kechida et al [34].	2004 2011	39,1%	29,5	9,6	28,8
Horton et al [45].	1999 2011	30,8%	19,5%	11,3%	-
Koonarat et al [38]	2010 2012	15,1%	12,7%	2,4%	-
Our series	2013	27.9%	21.3%	5.1%	1.4%

	2017				

2.7.2.2. Constitutional thrombophilia

In our series, CT was found in 207 patients or 21.3% of whom 184 patients had a single constitutional defect (18.9%) and 24 patients had combined constitutional defects (2.4%).

Our results are similar to those of the Tunisian study by Ben Salah et al[32] which revealed a frequency of constitutional thrombophilia of the order of 22.6%.

In most of the studies cited above, constitutional thrombophilia is much more common than acquired thrombophilia in the same study population, which is consistent with our findings. **Table XII** illustrates the frequency of the main constitutional thrombophilia anomalies in different series.

Table XII: Comparative frequencies of the main constitutional thrombophilia anomalies in different series in the literature

	Deficit in PS	PC deficit	TA deficit	CPRa
Frequency in the general population[46, 47]	0,03-0,13%	0,14-0,5%	0,02-0,17%	-
Frequency in patients with VTE[46, 48]	1,4-7,5%	1,4-8,6%	0,5-4,9%	-
Kechida et al [34].	28,8%	7,7%	3,8%	51,9%
Hentati et al [48]	4%	7%	-	2%
Turan et al [49].	13,5%	5,7%	1,1%	-
Koonarat et al. [33]	3%	1,2%	0,6%	-
Our series	**3,9%**	**2,7%**	**4,5%**	**13,6%**

PS deficiency was found in 38 patients (3.9%). Our results are in agreement with the literature, in particular the Tunisian study by Hentati et al **[48]** which found a frequency of 4%.

PC deficiency was reported in 28 patients or 2.8%, which is not in agreement with the results of the previously cited studies where this frequency varied from 1.2% **[33]** to 8.6% **[46]**. However, our result is clearly lower than those revealed by the two Tunisian studies where the frequency of PC deficiency was 7.7% in the study by Kechida et al [34] and 7% in the study by hentati et al [48]. Otherwise, AT deficiency was detected in 44 of our patients with a frequency of 4.5% which is in agreement with the results of some studies cited previously [33].

On the other hand, PCaR was detected in 138 of our patients (14.2%), thus presenting the most frequent constitutional thrombophilia anomaly in our study series. Our results are in agreement with the Tunisian study by Kechida et al [34] who found that cPTR was the most common constitutional thrombophilia in their study population.

Indeed, cPDR in most cases reflects a mutation in FVL affecting one of the cleavage sites of factor V by PCa resulting in deficient inactivation of this factor [50].

According to the results of the molecular biology, which only concerned the thrombophilia check-ups of the year 2017, the FVL mutation was found to be responsible for 36/36 of the CPPa. The mutation was heterozygous in 34 patients and homozygous in the remaining two patients.

Several studies have shown that the FVL mutation is a frequent mutation in the Sfax region, notably the study by Maalej et al [51] on blood donors which showed that this mutation existed in 13.6% of the study population with a predominance of the heterozygous profile and this is explained by several authors by the high rate of endogamy and consanguineous management in this region.

2.7.2.3.   Acquired thrombophilia

The frequency of acquired thrombophilia in patients with VTE varies in the literature from 2.4% [33] to 19.1% [32].

In our series, 64 patients had a positive CCA type LA, i.e. 6.6% of our population, with a predominance of females. This female predominance was also found in the study by Zhao et al [52].

The same study revealed a frequency of thromboembolic events of the order of

67% in these patients, which is in line with our results. Indeed, a venous thromboembolic event was found in the majority of our patients with LA CCA.

The mean age of our patients with LA CCA was 39.6 years, which is close to the results provided by the Chinese study by Shi et al [53] who found a mean age of 41 years.

2.7.2.4. Mixed thrombophilia

The frequency of mixed thrombophilia in our population was 1.4%, which was the presence of LA type ACC with a natural coagulation inhibitor deficiency or with CPPa. The latter was the most frequent case with a frequency of 0.8%. Our results are similar to those of Mateo et al [44] who found a mixed thrombophilia associating an LPA with a TC in 0.75% of patients.

2.7.2.5. In arterial thrombosis

In our series, constitutional thrombophilia was found in 55 patients or 26%. Isolated constitutional thrombophilia represented 14.6% while combined constitutional thrombophilia presented 11.3%.

A type LA ACC was found in isolation in 24 patients or 11.3%. This frequency is similar to that found in the study by Spagnoli et al [54]. Mixed thrombophilia was found in seven patients with a frequency of 3.3%.

The following **table** illustrates the frequency of thrombophilia abnormalities in some of the literature series.

Table XIII: Comparative frequencies of thrombophilia in BP in selected literature series

Study	DeficitType	Deficit TA	Deficit in PS	RPCa in PC	ACCtype	AT THE
Carod et al[55].	STROKE	11,5%	0,76%	0%	2,3%	0,5%
Spagnoli et al[54].	IDM	13%	5%	5,5%	-	11%
Naziha et	STROKE	16,67%	16,67%	-	-	-

al[56].						
Stepien et	STROKE	2,4%	1,2%	3,6%	-	-
al[57].	IDM	2,4%	2,4%	1,2%	-	-
Our study	**TA**	**5,1%**	**2,3%**	**2,3%**	**12,7%**	**14,6%**
	STROKE	4,7%	2,3%	1,4%	9,4%	8%
	IDM	0,4%	0%	0,4%	2,3%	3,7%

In our series, a PS deficit was found in ten stroke patients, with a frequency of 4.7%. This frequency is close to the results of the previously cited studies. Otherwise, the frequency of CPPa in stroke patients was 9.4%, which is higher than the frequency found in the study by Carod et al **[55]**, which can be explained by the high rate of CPPa in the Sfax region.

Thrombophilia is a multigenic disorder that requires a coherent, comprehensive and controlled thrombophilia work-up, taking into account the epidemiological, clinical and environmental data of patients with thrombotic events in order to establish a reliable interpretation of the work-up.

Thus, the aim of our retrospective study was to evaluate the requests for thrombophilia assessments, to report the epidemiological, topographical and etiological characteristics of thrombotic accidents and to determine the biological profile of thrombophilia assessment during these accidents in subjects aged between 18 and 60 years.

Based on the GFEHT recommendations detailing the indications for thrombophilia testing in the context of VTE, not all requests in our study were relevant. Only 423 (43.6%) requests were justified. On the other hand, the indication for thrombophilia testing in the context of TA remains controversial in the literature.

In our study, 128 tests were complete (10.8%) and 32 tests were controlled (2.7%). In fact, the non-exhaustiveness of the tests prevented us from determining the exact frequency of thrombophilia anomalies found in our study population.

In patients with VTE, the mean age was 40.64 years with a predominance of females, while in patients with BP, the mean age was 36.7 years with a predominance of males.

For patients with VTE, constitutional thrombophilia was found in 208 patients or 21.4%, of which 184 patients had a single abnormality and 24 patients had a combined abnormality. The LA type ACC was found in isolation in 50 patients or 5.1% and in association with constitutional thrombophilia in 14 patients or 1.4%.

PS deficiency was observed in 38 patients (3.9%), PC deficiency was also observed in 28 patients (2.8%), AT deficiency was observed in 44 of our patients (4.5%) and CPPa was found in 138 of our patients (14.2%) thus presenting the most common abnormality encountered in our study

For patients with BP, isolated constitutional thrombophilia was found in 31 patients (14.6%) and combined constitutional thrombophilia was found in 4 patients (1.8%).

LA CCA was found in isolation in 24 patients (11.3%) and in association with constitutional thrombophilia in 7 patients (3.3%).

PS deficiency was observed in 11 patients (5.2%), PC deficiency was observed in 5 patients (2.3%), AT deficiency was observed in 5 patients (2.3%) and CPPa was found in 27 patients (12.7%), thus presenting the most common anomaly in our study population.

Taking into consideration the high cost of thrombophilia assessment, a review of the indications for this assessment is necessary in order to restrict its use and to obtain more relevant and targeted research.

Among the limitations of our etiological exploration:

- The FVL transfer was only made for the 2017 balance sheets.

- The prothrombin mutation G202120A was not tested.

- The search for LPAs other than LA type ACCs has not been done.

In conclusion, our study has allowed us to emphasise the importance of clinical information, comprehensive investigations of thrombophilia markers and, above all, the biological control of results and to recall that cooperation between the biologist and the clinician is the best way to obtain relevant, targeted and reliable results.

BIBLIOGRAPHIC REFERENCES

1. Sanchez O, Benhamou Y, Bertoletti L, Constant J, Couturaud F, Delluc A, et al. Good practice recommendations for the management of venous thromboembolic disease in adults. Short version. Rev Mal Respir. 2019;36(2):249-83.

2. Carroll BJ, Piazza G. Hypercoagulable states in arterial and venous thrombosis: When, how, and who to test? Vasc Med. 2018;23(4):388-99.

3. Samaher B. Role of protein C, a natural anticoagulant, in thrombosis and cancer association [These]. Monastir: Faculty of Pharmacy of Monastir; 2015.

4. Bijak M, Rzeznicka P, Saluk J, Nowak P. Cellular model of blood coagulation process. Pol Merkur Lekarski. 2015;39(229):5-8.

5. Smock KJ, Plumhoff EA, Meijer P, Hsu P, Zantek ND, Heikal NM, et al. Protein S testing in patients with protein S deficiency, factor V leiden, and rivaroxaban by north American specialized coagulation laboratories. Thromb Haemost. 2016;116(1):50-7.

6. Deng MY, Liu ZX, Huang HF, Chen YH, Luo YJ, Sun NN, et al. Two novel compound heterozygous mutations associated with types I and II protein C deficiency with unusual phenotypes. Thromb Res. 2016;145:93-9.

7. Dinarvand P, Moser KA. Protein C deficiency. Arch Pathol Lab Med. 2019;143(10):1281-5.

8. Dinarvand P, Hassanian SM, Weiler H, Rezaie AR. Intraperitoneal administration of activated protein C prevents postsurgical adhesion band formation. Blood. 2015;125(8):1339-48.

9. Liu H, Wang HF, Tang L, Yang Y, Wang QY, Zeng W, et al. Compound heterozygous protein C deficiency in a family with venous thrombosis: Identification and in vitro study of p.Asp297His and p.Val420Leu mutations. Gene. 2015;563(1):35-40.

10. Aguila S, Martinez-Martinez I, Dichiara G, Gutierrez-Gallego R, Navarro-Fernandez J, Vicente V, et al. Increased N-glycosylation efficiency by generation of an aromatic sequon on N135 of antithrombin. PLoS One.

2014;9(12):e114454.

11. Bauer KA, Nguyen-Cao TM, Spears JB. Issues in the diagnosis and management of hereditary antithrombin deficiency. Ann Pharmacother. 2016;50(9):758-67.

12. Corral J, De la Morena-Barrio ME, Vicente V. The genetics of antithrombin. Thromb Res. 2018;169:23-9.

13. Mulder R, Croles FN, Mulder AB, Huntington JA, Meijer K, Lukens MV. SERPINC1 gene mutations in antithrombin deficiency. Br J Haematol. 2017;178(2):279-85.

14. Lam W, Moosavi L. Physiology, Factor V [Online]. 2019 [accessed 03/09/2019]. Available from: https://www.ncbi.nlm.nih.gov/books/ NBK544237/

15. Patel K, Fasanya A, Yadam S, Joshi AA, Singh AC, DuMont T. Pathogenesis and epidemiology of venous thromboembolic disease. Crit Care Nurs. 2017;40(3):191-200.

16. Favaloro EJ. Genetic testing for thrombophilia-related genes: observations of testing patterns for factor V Leiden (G1691A) and prothrombin gene "Mutation" (G20210A). Semin Thromb Hemost. 2019;45(7):730-42.

17. Dahlback B. Pro- and anticoagulant properties of factor V in pathogenesis of thrombosis and bleeding disorders. Int J Lab Hematol. 2016;38:4-11.

18. Izuhara M, Shinozawa K, Kitaori T, Katano K, Ozaki Y, Fukutake K, et al. Genotyping analysis of the factor V Nara mutation, Hong Kong mutation, and 16 single-nucleotide polymorphisms, including the R2 haplotype, and the involvement of factor V activity in patients with recurrent miscarriage. Blood Coagul Fibrinolysis. 2017;28(4):323-8.

19. Pengo V, Bison E, Banzato A, Zoppellaro G, Jose SP, Denas G. Lupus anticoagulant testing: diluted russell viper venom time (dRVVT). Methods Mol Biol. 2017;1646:169-76.

20. Molhoek JE, De Groot PG, Urbanus RT. The lupus anticoagulant paradox. Semin Thromb Hemost. 2018;44(5):445-52.

21. Joste V, Dragon-Durey MA, Darnige L. Laboratory diagnosis of antiphospholipid syndrome: From criteria to practice. Rev Med Interne. 2018;39(1):34-41.

22. Tran HA, Gibbs H, Merriman E, Curnow JL, Young L, Bennett A, et al. New guidelines from the Thrombosis and Haemostasis Society of Australia and New Zealand for the diagnosis and management of venous thromboembolism. Med J Aust. 2019;210(5):227-35.

23. Spychalska-Zwolinska M, Zwolinski T, Mieczkowski A, Budzynski J. Thrombophilia diagnosis: a retrospective analysis of a single-center experience. Blood Coagul Fibrinolysis. 2015;26(6):649-54.

24. Kushner A, West DW, Pillarisetty LS. Virchow triad [Online]. 2019 [accessed 11/11/2019]. Available from: https://www.ncbi.nlm.nih.gov/books/NBK539697/

25. Allain JS, Gueret P, Le Gallou T, Cazalets C, Lescoat A, Jego P. Hereditary thrombophilia testing and its therapeutic impact on venous thromboembolism disease: Results from a retrospective single-center study of 162 patients. Rev Med Interne. 2016;37(10):661-6.

26. Kwang H, Mou E, Richman I, Kumar A, Berube C, Kaimal R, et al. Thrombophilia testing in the inpatient setting: impact of an educational intervention. BMC Med Inform Decis Mak. 2019;19(1):167-9.

27. Gavva C, Johnson M, De Simone N, Sarode R. An audit of thrombophilia testing in patients with ischemic stroke or transient ischemic attack: the futility of testing. J Stroke Cerebrovasc Dis. 2018;27(11):3301-5.

28. Devignes J, Smail-Tabbone M, Herve A, Cagninacci G, Devignes MD, Lecompte T, et al. Extended persistence of antiphospholipid antibodies beyond the 12-week time interval: Association with baseline antiphospholipid antibodies titres. Int J Lab Hematol. 2019;41(6):726-30.

29. Journaud M, Ferreira-Maldent N, Gruart VG, Maillot F. Evaluation of antiphospholipid antibody testing at the CHRU of Tours: one-year retrospective study. Rev Med Interne. 2017;38:182-3.

30. Yang JY, Chan AK. Pediatric thrombophilia. Pediatr Clin North Am. 2013;60(6):1443-62.

31. Connors JM. Thrombophilia testing and venous thrombosis. N Engl J Med. 2017;377(12):1177-87.

32. Ben Salah R, Frikha F, Kaddour N, Saidi N, Snoussi M, Marzouk S, et al. Risk factor for deep venous thrombosis in internal medicine: A retrospective study of 318 cases. Ann Cardiol Angeiol. 2014;63(1):11-6.

33. Koonarat A, Rattarittamrong E, Tantiworawit A, Rattanathammethee T, Hantrakool S, Chai-Adisaksopha C, et al. Clinical characteristics, risk factors, and outcomes of usual and unusual site venous thromboembolism. Blood Coagul Fibrinolysis. 2018;29(1):12-8.

34. Kechida M, Nasr MB, Klii R, Marzouk M, Hammami S, Khochtali I. Thrombosis due to thrombophilia anomalies: what particularities? Rev Med Interne. 2016;37:150-3.

35. Martinez C, Cohen AT, Bamber L, Rietbrock S. Epidemiology of first and recurrent venous thromboembolism: a population-based cohort study in patients without active cancer. Thromb Haemost. 2014;112(2):255-63.

36. Faioni EM, Zighetti ML, Vozzo NP. Sex, gender and venous thromboembolism: do we care enough? Blood Coagul Fibrinolysis. 2018;29(8):663-7.

37. Konin C, Soya E, Yapi B, Monney E, Ekou A, N'djessan JJ, et al. Common denominators between venous and arterial thrombosis in a sub-Saharan African population. J Med Vasc. 2018;43(2):129.

38. Suchon P, Resseguier N, Ibrahim M, Robin A, Venton G, Barthet MC, et al. Common risk factors add to inherited thrombophilia to predict venous thromboembolism risk in families. Thromb Haemost Open. 2019;3(1):28- 35.

39. Kefi A, Larbi T, Abdallah M, Ouni AE, Bougacha N, Bouslama K, et al. Young ischemic stroke in Tunisia: a multicentric study. Int J Neurosci. 2017;127(4):314-9.

40. Guder WG. History of the preanalytical phase: a personal view. Biochem

Med. 2014;24(1):25-30.

41. Magnette A, Chatelain M, Chatelain B, Ten Cate H, Mullier F. Pre-analytical issues in the haemostasis laboratory: guidance for the clinical laboratories. Thromb J. 2016;14:49-63.

42. Lipets EN, Ataullakhanov FI. Global assays of hemostasis in the diagnostics of hypercoagulation and evaluation of thrombosis risk. Thromb J. 2015;13(1):4-18.

43. Lim MY, Moll S. Thrombophilia. Vasc Med. 2015;20(2):193-6.

44. Mateo J, Oliver A, Borrell M, Sala N, Fontcuberta J. Laboratory evaluation and clinical characteristics of 2,132 consecutive unselected patients with venousthromboembolism--results of   the Spanish Multicentric Study on Thrombophilia (EMET-Study). Thromb Haemost. 1997;77(3):444-51.

45. Garcia-Horton A, Kovacs MJ, Abdulrehman J, Taylor JE, Sharma S, Lazo-Langner A. Impact of thrombophilia screening on venous thromboembolism management practices. Thromb Res. 2017;149:76-80.

46. Kyrle PA, Eichinger S. Deep vein thrombosis. Lancet. 2005;365(9465):1163-74.

47. Puhr HC, Eischer L, Sinkovec H, Traby L, Kyrle PA, Eichinger S. Circumstances of provoked recurrent venous thromboembolism: the Austrian study on recurrent venous thromboembolism. J Thromb Thrombolysis. 2019. In Press. doi: 10.1007/s11239-019-01965-z.

48. Hentati O, Jaziri F, Skouri W, Bennaser M, Ben AK, Sami T. Topographical, etiological and evolutionary profile of deep vein thrombosis in young subjects in an internal medicine department. Rev Med Interne. 2017;38:115-6.

49. Turan O, Undar B, Gunay T, Akkoclu A. Investigation of inherited thrombophilias in patients with pulmonary embolism. Blood Coagul Fibrinolysis. 2013;24(2):140-9.

50. Amiral J, Vissac AM, Seghatchian J. Laboratory assessment of Activated

Protein C Resistance/Factor V-Leiden and performance characteristics of a new quantitative assay. Transfus Apher Sci. 2017;56(6):906-13.

51. Maalej L, Hadjkacem B, Ben Amor I, Smaoui M, Gargouri A, Gargouri J. Prevalence of factor V Leiden in south Tunisian blood donors. J Thromb Thrombolysis. 2011;32(1):116-9.

52. Zhao JL, Sun YD, Zhang Y, Xu D, Wang Q, Li MT, et al. The clinical manifestations and thrombotic risk factors in primary antiphospholipid syndrome. Zhonghua Nei Ke Za Zhi. 2016;55(5):386-91.

53. Shi H, Teng JL, Sun Y, Wu XY, Hu QY, Liu HL, et al. Clinical characteristics and laboratory findings of 252 Chinese patients with antiphospholipid syndrome: comparison with Euro-Phospholipid cohort. Clin Rheumatol. 2017;36(3):599-608.

54. Spagnoli V, Diefenbronn M, Merat B, Logeart D, Sideris G, Henry P, et al. ST elevation myocardial infarction in young adults: Is there an interest for thrombophilia screening? Ann Cardiol Angeiol. 2019;68(2):98-106.

55. Carod-Artal FJ, Nunes SV, Portugal D, Silva TV, Vargas AP. Ischemic stroke subtypes and thrombophilia in young and elderly Brazilian stroke patients admitted to a rehabilitation hospital. Stroke. 2005;36(9):2012-4.

56. Khammassi N, Sassi YB, Aloui A, Kort Y, Abdelhedi H, Cherif O. Ischemic stroke in young patients: about 6 cases. Pan Afr Med J. 2015;22:142-6.

57. Stepien K, Nowak K, Wypasek E, Zalewski J, Undas A. High prevalence of inherited thrombophilia and antiphospholipid syndrome in myocardial infarction with non-obstructive coronary arteries: Comparison with cryptogenic stroke. Int J Cardiol. 2019;290:1-6.

Appendix 1: Profile of thrombophilia in the context of venous thrombosis

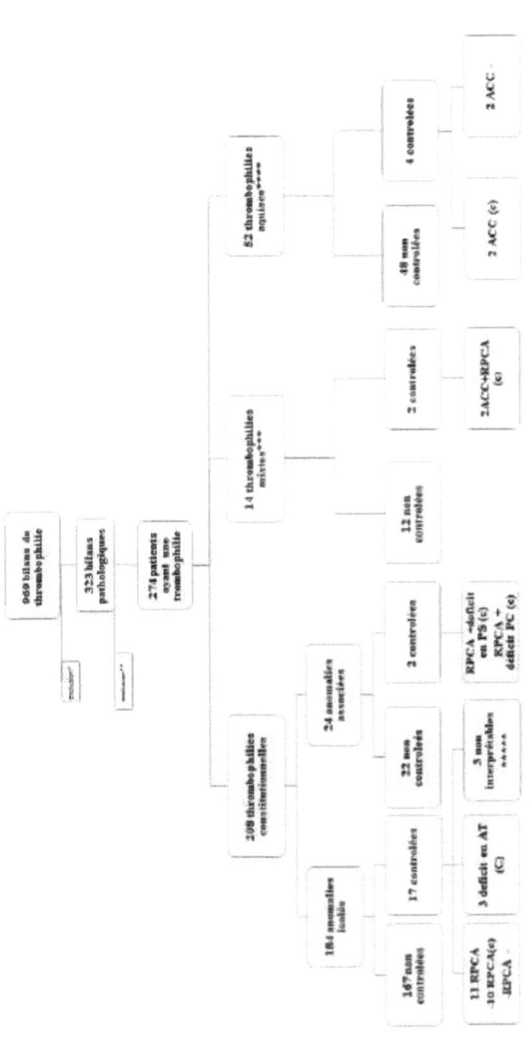

Appendix 2: Profile of thrombophilia in arterial thrombosis

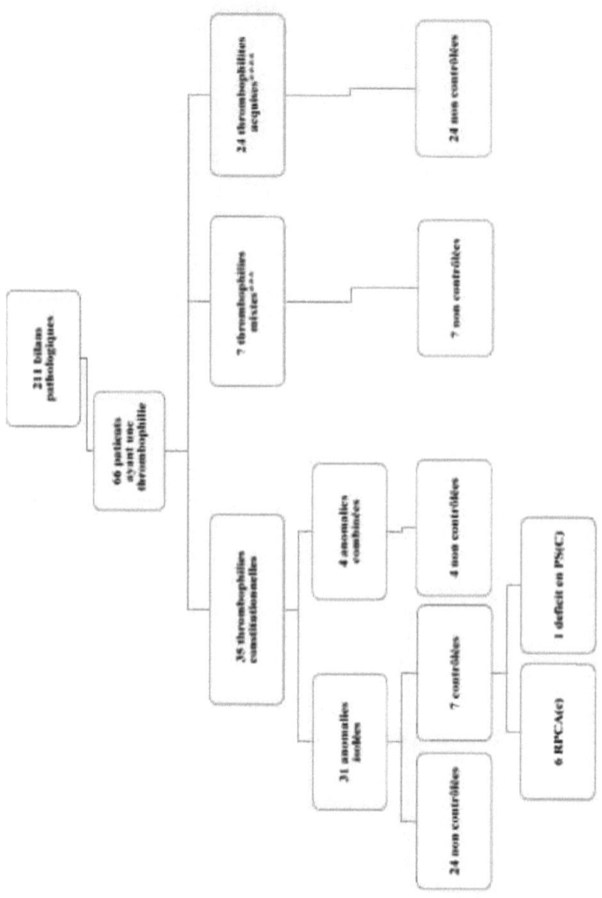

Printed by Books on Demand GmbH, Norderstedt / Germany